Our Exciting
PONDS

By Jagger Youssef

Please visit our website, www.garethstevens.com. For a free color catalog of all our high-quality books, call toll free 1-800-542-2595 or fax 1-877-542-2596.

Cataloging-in-Publication Data

Names: Youssef, Jagger.
Title: Ponds / Jagger Youssef.
Description: New York : Gareth Stevens Publishing, 2018. | Series: Our exciting Earth! | Includes index.
Identifiers: ISBN 9781538209738 (pbk.) | ISBN 9781538209752 (library bound) | ISBN 9781538209745 (6 pack)
Subjects: LCSH: Ponds–Juvenile literature. | Pond ecology–Juvenile literature.
Classification: LCC QH541.5.P63 Y68 2018 | DDC 577.63'6–dc23

Published in 2018 by
Gareth Stevens Publishing
111 East 14th Street, Suite 349
New York, NY 10003

Copyright © 2018 Gareth Stevens Publishing

Editor: Therese Shea
Designer: Bethany Perl

Photo credits: Cover, p. 1 kavram/Shutterstock.com; p. 5 Jacek Boczarski/Shutterstock.com; p. 7 MH Anderson Photography/Shutterstock.com; p. 9 Lucy/Shutterstock.com; p. 11 Etalita/Shutterstock.com; p. 13 S.Z./Shutterstock.com; p. 15 pedphoto36pm/Shutterstock.com; p. 17 Matteo photos/Shutterstock.com; p. 19 Phillip B. Espinasse/Shutterstock.com; p. 21 Erni/Shutterstock.com; p. 23 Dimedrol68/Shutterstock.com.

All rights reserved. No part of this book may be reproduced in any form without permission in writing from the publisher, except by a reviewer.

Printed in the United States of America

CPSIA compliance information: Batch #CW18GS: For further information contact Gareth Stevens, New York, New York at 1-800-542-2595.

Contents

Visit a Pond! 4

Plenty of Plants 8

A Lot of Algae 12

Animal Home 14

Finding Fish 20

Words to Know 24

Index 24

Let's go to a pond!

A pond is smaller than a lake.
It's not deep.
It has freshwater.

Plants live around ponds. Some live in ponds!

I see a water lily.
Its roots are
at the pond's bottom.
Its leaves are on top!

Algae look like plants.
They can make
a pond green!

A pond is a home for animals.
There are many places to hide!

Frogs live near ponds.
They lay eggs
in the water!

Ducks live near ponds.
I see ducklings swimming!

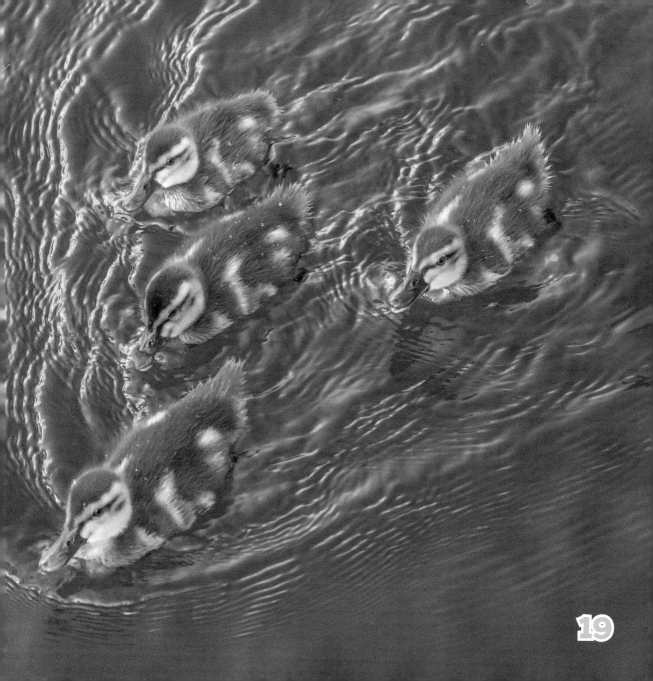

Fish live in ponds.
Minnows eat
bugs and algae!

We look for fish.
Ponds are fun!

Words to Know

 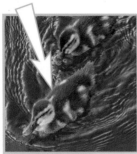  

algae   duckling   minnow   water lily

Index

algae   12, 20   fish   20, 22

ducks   18   frogs   16